The Health and Safety Handbook

The Health and Safety Handbook

A one-stop guide for managers

by

Pat McGuinness and Lynn Smith

Revised edition
First published in 2004 by
Spiro Press
17-19 Rochester Road
London SW1P 1LA
Telephone: 0870 400 1000

ISBN 1 84439 016 0

British Library Cataloguing-in-Publication Data.
A catalogue record for this book is available from the British Library.

Typeset by: The Midlands Book Typesetting Company and
J W Arrowsmith Ltd
Printed by: J W Arrowsmith Ltd
Cover design: Compendium

Contents

Foreword

This book has been designed to help those who manage in organisations to unravel their health and safety at work responsibilities and provide some practical ideas for making a success of their management. It may well provide some answers, but we also hope that it will prompt the reader to ask quality questions regarding the management of health and safety.

Health and safety at work is often seen as a burden, as an issue that gets in the way of production and service. The reality is, of course, that a well-managed organisation has few problems in managing its health and safety requirements.

In our experience, organisations that are struggling with health and safety are likely to be struggling with other areas of their business also.

Authors' note

While every effort has been made to ensure the accuracy of the information in this book, legislation is constantly under review and no legal responsibility can be accepted by Spiro Press. Advice in this book is intended as guidance only. Readers should seek professional advice for specific, personal issues.

All instances of the Health and Safety at Work Act 1974 should read the Health and Safety at Work etc Act 1974.

An explanation of the symbols used in this book

💣 This symbol is used to denote something you should attend to without delay, for rather like an unexploded bomb if you don't take any action to defuse it, the consequences could be serious for both you and your organisation!

✎ This symbol is used to denote a hint or tip that will help you with your health and safety management.

Introduction

Managers, supervisors, team leaders and section heads all have to cope with implementing the employer's duties under health and safety at work legislation. Like everyone else in the organisation they also have their health and safety duties as employees to deal with. This 'meat in the sandwich' role is often seen as daunting.

What is being faced here is the turning of policy and procedures into practice. This, of course, is not just a health and safety issue! However, whilst this may all sound difficult, it does not need to be. By getting some basic awareness of what is required, learning from others and knowing where to obtain advice, a great deal can be achieved.

Effective health and safety management is a combination of common sense, vigilance and rigour. In most accident situations analysis shows that either fairly simple preventive and/or corrective measures or the rigorous application of some existing system would have prevented the incident.

This book will review current legislation and the associated Approved Codes of Practice (ACOPs) or Guidance. Providing practical advice on how to meet the requirements of the legislation and manage health and safety in a cost-effective manner, the book aims to help promote a positive health, safety and welfare culture.

CHAPTER 1

Historical background

The history of health and safety legislation can be traced back to the passing of the first factory statute in 1802, enacted as a result of concern about the appalling working conditions of children in the cotton mills.

The next 170 years saw the piecemeal extension of legislation to cover other types of workers, workplaces and hazards, including the Factories Act 1961 and the Offices, Shops and Railway Premises Act 1963, some parts of which are still in force today.

In 1972 the Robens Report (chaired by Lord Robens) underlined that the primary responsibility for high levels of occupational accidents and disease lay with those who create the risks and those who work with them. The Report formed the basis of the legal framework of the Health and Safety at Work etc Act 1974.

The 1974 Act itself is the most important plank of a number of pieces of legislation concerned with the health, safety and welfare of people. It imposed legal duties on both employers and employees (including those in a management role), and established a single, comprehensive, integrated system of law.

CHAPTER 2

Health and safety legislation – an overview

Health and Safety at Work etc Act 1974

This Act represents the principal piece of health and safety legislation in the UK, under which all other Regulations are made.

The new Act did not repeal or replace existing legislation relating to specific occupations and workplaces, but covered all categories of employment (including the self-employed) and dealt with the general duties and obligations of all workers by encouraging proper safety management and enforcement.

The Act also provides the power for the relevant government minister to draw up a system of Regulations and ACOPs that would progressively replace existing legislation, with the aim of continually improving the standard of health, safety and welfare. As such, the Health and Safety at Work etc Act is known as an 'enabling' Act, being a generalised piece of legislation that 'enables' the introduction of more detailed and specific legislation.

The Act also set up the Health and Safety Commission (HSC), a body consisting of a chairperson and not fewer than six, or more than nine, other members. Of the members, three represent the employers and three the employees (trade unions); the remainder are drawn from other bodies such as local authorities and professional organisations. The HSC has overall responsibility for administering and overseeing health and safety matters, including the effective operation of health and safety law. It also can submit proposals for new Regulations and replace/ update existing legislation.

The Act also established the Health and Safety Executive (HSE), the operational arm of the HSC. We explain the role of the HSE later in this chapter.

Regulations, Approved Codes of Practice (ACOPs) and Guidance Notes

The **Regulations** have the force of law and render anyone in breach of them liable to prosecution. Regulations are often made as a result of European Union (EU) Directives on specific areas of health and safety.

Approved Codes of Practice (ACOPs) are not law in themselves but have persuasive authority and provide practical guidance on health and safety matters; they interpret an Act or Regulation.

Guidance Notes are often produced by the HSE. They may be referred to by the HSE inspectors to illustrate what they regard as minimum standards. In any prosecution where it is established that the relevant provisions of the Code have not been followed, the defendant must show that the law has been complied with in some other way.

Legislation from the European Union

Under the Single European Act the member states agreed that health and safety legislation should be harmonised throughout the EU, with an emphasis on the need to improve the working environment. The working environment, in the EU's view, encompassed such considerations as ergonomics (the study of the relationship between people and their environment) as well as the traditional health and safety measures. The strategy of the EU is to use Directives, which apply the minimum standards with which the law in each member state must comply. There can be, however, different ways of meeting these requirements, and the responsibility for how it is achieved is left with each individual state.

EU Directives on health and safety are based on a set of common principles and themes which are likely to follow the same format in the future.

- The avoidance of risks.
- The evaluation of risks which cannot be avoided.
- The necessity to combat and deal with risk at source.
- The replacement of the dangerous by the non-dangerous or the less dangerous.
- The need to give collective protective measures priority over individual protective measures.

In 1992 six major pieces of European legislation came into being. These became known in the UK as the '**six pack**'.

The practical impact of this new legislation was not to add more requirements, but rather to make explicit what was already implicit in the Health and Safety at Work etc Act.

The 'six pack' consists of:

- Management of Health and Safety at Work Regulations (amended 1999)
- Workplace (Health, Safety and Welfare) Regulations 1992
- Provision and Use of Work Equipment Regulations 1992 (PUWER) (updated 1998 - PUWER 98)
- Manual Handling Operations Regulations 1992
- Personal Protective Equipment at Work Regulations 1992 (PPE)
- Health and Safety (Display Screen Equipment) Regulations 1992 (DSE).

✎ *You are strongly advised to obtain from the HSE their six booklets specifically on these Regulations. They contain not only the Regulations but also ACOPs and Guidance on how to comply. Chapter 3 outlines these Regulations in more detail.*

The legal system in operation

The phrase 'so far as is reasonably practicable' is key within the Health and Safety at Work etc Act. For example, employers must weigh up the costs of providing a safe system of work against the risk to health and safety if such a system is not provided. Only if the costs are grossly disproportionate to the risks can the safety precaution be considered unreasonable. Most everyday health and safety issues, however, will have 'reasonably practicable' solutions. The cost of keeping walkways and fire exits clear, for example, is not disproportionately high compared with the risk to safety.

The thinking behind this comes from a legal case, *Edwards* v *NCB* [1949], where the judge stated: 'A computation must be made in which the quantum of risk is placed on one scale, and the sacrifice involved in the measure for averting the risk is placed on the other'.

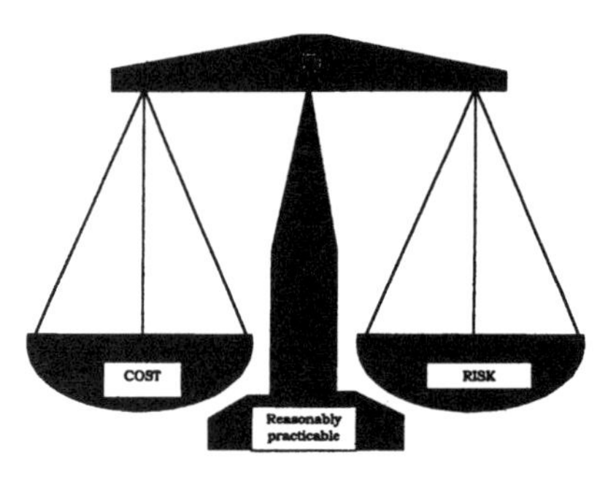

Figure 1. 'Reasonably practicable' – the balance of cost and risk.

When deciding 'cost v risk', the size of the risk and the sacrifice needed to reduce the risk, two factors must also be looked at:

1. The likelihood of injury being in fact caused (the greater the probability that an accident will occur, the greater the duty to guard against it).
2. The severity of injury risked.

> *Within the legislation you will come upon various words which qualify your legal duties:* ***absolute*** *– you have to do this (there is no choice);* ***practicable*** *– if it is feasible to do;* ***reasonably practicable*** *– cost v risk.*

The HSE

The Health and Safety at Work etc Act, as we have seen, also brought into being the HSE, an enforcement and policy-development organisation. Day-to-day enforcement is undertaken by two bodies:

1. The HSE Inspectorate.
2. Local authority environmental health officers.

In general, the HSE are responsible for enforcement in industrial premises and the local authorities are responsible for enforcement in offices and shops. Where offices are part of factories, hospitals, etc. the enforcement body would be the HSE.

Health and safety enforcement officers

Health and safety enforcement officers have wide powers to enter premises, test materials and conduct investigations – *seize, destroy or render harmless*. When an enforcement officer visits a place of work, either on a routine or special visit, the aim is to ensure the observance, maintenance and improvement of health and safety standards, and in doing so, to secure compliance with the law. A wide variety of procedures are open to the inspector if any action is required:

Informal procedures

- verbal advice
- letters and provision of information by means of advisory leaflets, etc.

Formal procedures

- improvement notices
- prohibition notices
- prosecution.

✎ *Get to know your local inspector; they can be very helpful and give good advice.*

💣 *Do not deny entry to an enforcement officer!*

💣 *You don't have to have had an accident to be prosecuted for breaching health and safety law.*

During a visit an enforcement officer may discuss any necessary improvements with interested parties, including managers, safety representatives and employees, as appropriate. In deciding whether to give verbal or written advice, issue notices or consider legal proceedings, an enforcement officer will consider various factors:

- the seriousness of the risk
- any breach of legislation
- the methods available for remedying the situation
- general standards in the organisation, including its health and safety policy and procedures and their implementation.

The enforcement officer will then decide which form of action to take.

Improvement and prohibition notices

Improvement notices

Where an organisation or an individual contravenes the legislation, an enforcement officer may issue an 'improvement notice' requiring the organisation and/or the individual to improve the situation. The enforcement officer *may* also give a letter of intent prior to the notice warning of its arrival.

Key elements of an improvement notice:

- issued for breach of the law
- minimum of 21 days to comply
- can appeal
- notice will be suspended until appeal heard.

> *Treat improvement notices as if they are prohibition notices and do something quickly about the remedial action. Think! What would the consequences be if an accident occurred whilst the activity/equipment was subject to an improvement notice?*

Prohibition notices

If an enforcement officer believes that a particular activity will cause serious risk or personal injury, a 'prohibition notice' may be served specifying remedial measures to be taken and by when. The prohibition notice may be deferred for a specified time or may take immediate effect, if the risk is imminent.

Key elements of a prohibition notice:

- issued where there is risk of serious injury
- 21 days to appeal
- effective immediately or by a specified date
- notice is not suspended whilst awaiting appeal
- can apply to equipment, building or organisation.

> ✎ *Whatever type of notice is issued, normally an enforcement officer will discuss and offer help to solve the problem – they want to help!*

Prosecution

An enforcement officer may, however, decide to prosecute for contravening legislation, failing to comply with an improvement or prohibition notice, or for committing any other specified offence, such as obstructing an enforcement officer or making false statements.

In making a decision the enforcement officer takes into account all the circumstances of the case, including its seriousness, the employer's record and the effectiveness of health and safety arrangements.

In most cases prosecutions are heard in magistrates courts (lower courts) where, if found guilty, fines of up to £20,000 per breach for sections 2–6 of the Health and Safety at Work etc Act 1974 can be awarded. For all other breaches of health and safety regulations, fines of up to £5,000 can be awarded. Also, a lower court can impose community service orders or imprison individuals for up to six months. Offences heard in the lower courts are known as 'summary offences'.

All of these options are regularly exercised by the lower courts. The fines can be substantial, because most incidents involve more than a single breach of legislation, so a form of 'totting up' takes place. These fines etc. can be awarded against the organisation and/or individuals within it.

You may, in a serious case, be indicted to crown court, where the penalties are more severe – unlimited fines and up to two years' imprisonment can be awarded.

News cuttings

- *You should ensure that you comply with your organisation's health and safety at work policies and procedures and the requirements of your job description.*
- *If you are responsible for the activities of others you should ensure that they comply with all policies and procedures relating to health and safety.*
- *If you are unfortunate enough to receive a personal prosecution your employer cannot pay the fine for you!*

✎ *Get it right first time.*

Effect of civil action (common law)

You should also bear in mind the increasing effect on organisations of civil action taken by employees or others who have suffered an injury or illness due to work activities – for example, *negligence*.

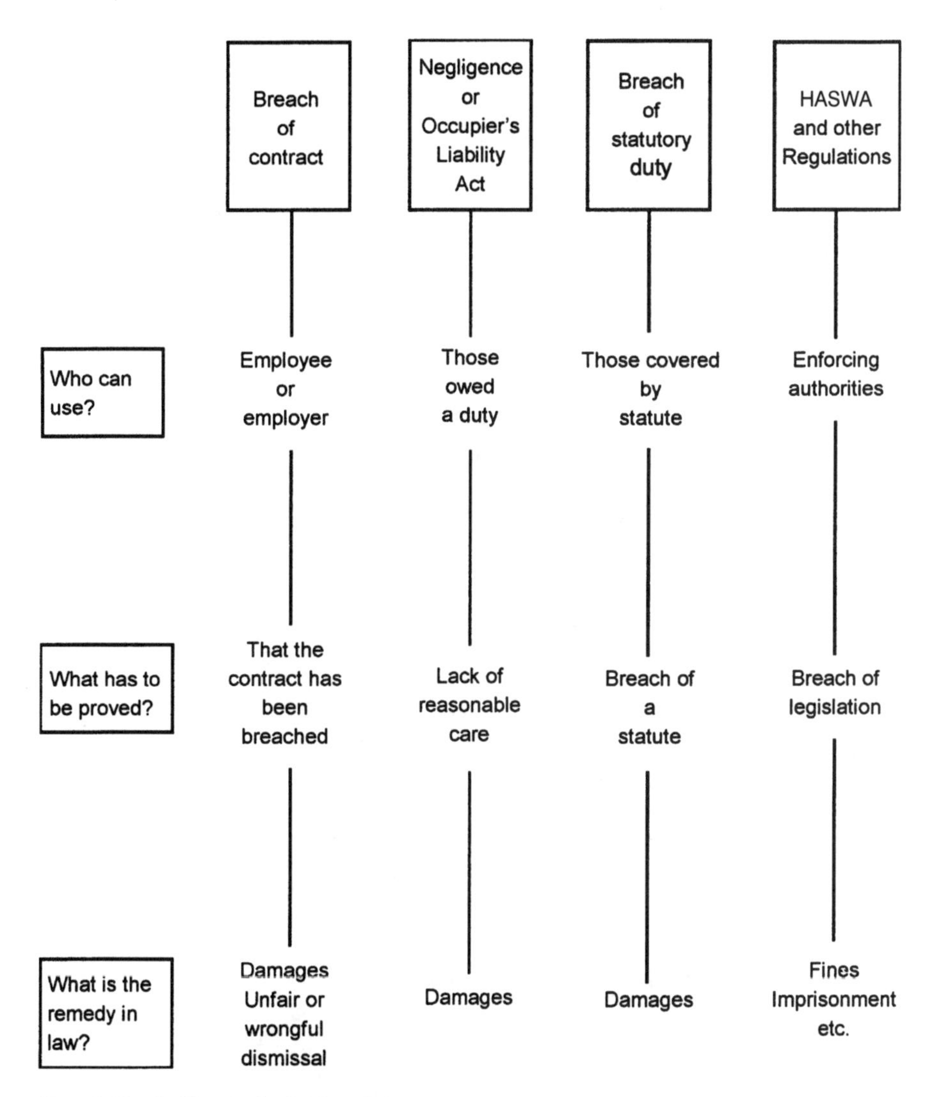

Figure 2. How health and safety law is applied.

The injured person, **the plaintiff**, brings an action of tort (a civil wrong as distinct from a criminal wrong) against **the defendant** – the employer. The plaintiff must prove that the employer breached his/her duty of care to the injured party.

There have been a number of highly publicised cases involving large sums of compensation to the individuals concerned. This has heightened the anxiety of insurers and led to a general rise in employers' insurance premiums.

> ✎ *In any accident situation make sure you note down the events that took place so that your insurers can have a full report of the incident.*
>
> ✎ *If you are able to demonstrate that you and the organisation have an effective health and safety management programme, it will not only help to protect everyone from harm but will also help to defend the organisation against any civil claims.*

News cuttings

CHAPTER 3

Health and safety legislation common to most workplaces

To understand fully the effect of any health and safety legislation, it is essential to have a clearer and more detailed understanding of the requirements and responsibilities of both employer and employee within the Health and Safety at Work etc Act 1974, and its subsequent Regulations, that are likely to affect most workplaces.

> ✎ *You should also bear in mind that some workplaces are governed by more specific Regulations where discrete/process activities take place such as:*
>
> - *Control of Asbestos at Work Regulations 2002*
> - *Gas Safety (Installation and Use) Regulations 1998*
> - *Construction (Design and Management) Regulations 2000*
> - *Food Safety (General Food Hygiene) Regulations 1995.*

A full list of current legislation is available from the HSE (address can be found in the Appendices).

You should as a general rule aim to comply with the requirements of the most specific Regulations that exist in any given situation. They usually have the highest and most absolute level of duty.

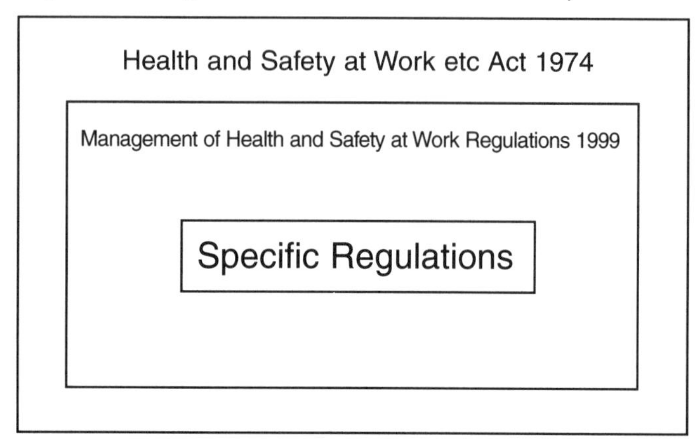

Figure 3. Specific Regulations

Any examination of UK health and safety legislation must start with the Health and Safety at Work etc Act 1974.

Health and Safety at Work etc Act 1974 (HASWA)

The Health and Safety at Work etc Act applies to all work situations. It covers everyone at work, whether they are employed or self-employed, and also protects anyone else whose health and safety may be affected by work activities – **the general public**. The Act also recognises that employees, as well as employers, have a duty to ensure high standards of health, safety and welfare at work.

General duties of employers

The Act imposes a duty on all employers to ensure, so far as is reasonably practicable, the health, safety and welfare at work of their employees. In particular this duty includes:

- providing and maintaining plant and systems of work which are safe and without risks to health
- making arrangements for ensuring safety and absence of risks to health in connection with the use, handling, storage and transport of articles and substances

- providing information, instruction, training and supervision to ensure the health and safety at work of all employees
- maintaining a workplace which is safe and without risks to health; providing and maintaining means of access to, and egress from, the workplace; providing adequate welfare facilities
- for employers who employ five people or more, preparing, and keeping up to date, a written statement of their policy showing how management intends to provide a safe working environment and also giving details of the organisational arrangements for carrying this out
- ensuring that non-employees (contractors, the general public, work experience staff, temporary staff, visitors, etc) are not exposed to any risk resulting from workplace activities
- providing free of charge to employees anything necessary, or required by law, in the interest of health and safety at work (personal protective equipment).

General duties of employees

The Act also acknowledges the importance of employees' involvement in health and safety at work. Employers are required to:

- take reasonable care for the health and safety of themselves and others who may be affected by their acts or omissions
- co-operate with employers and any other persons so far as is necessary to enable them to carry out statutory provisions
- not intentionally or recklessly interfere with or misuse anything provided in the interests of health, safety and welfare.

✎ *As a manager or designated leader, not only do you have to fulfil the responsibilities of the employer and of an employee but the Act also imposes extra duties:*

- *Where a person by his/her act or default causes another person to commit an offence, then he/she, as well as that other person, may be charged with the offence.*

- *Where an offence committed by a body corporate is shown to have been committed due to the consent, connivance or neglect of a director or senior manager, then he/she, as well as the body corporate, shall be guilty of the offence.*

One judge made the following comment in his summing-up of such a case: 'If there is evidence against the company there must be evidence against the directors'.

HASW Act in brief:

- Employer's duties to employees s.2
- Employer's duties to others s.3
- Employer/landlord s.4
- Employer/environment s.5
- Employer/articles and substances s.6
- Employee's duties to self and others s.7
- Employee not to interfere/misuse s.8
- Employer must not charge employees for the provision of health and safety equipment s.9
- Persons who let offence be committed s.36
- Consent, connivance of body corporate s.37

Management of Health and Safety at Work Regulations 1999

These Regulations set out more specific responsibilities for employers and employees, which are broadly similar to those in the HASWA. These requirements enable organisations to set an agenda to help them to '**manage**' health and safety effectively.

💣 *The duties are absolute – more use of phrases such as 'every employee shall' …*

✎ *This is your agenda for managing health and safety at work!*

General duties of employers

The main part of the Regulations stipulates that employers are required to assess potential hazards, record findings, identify preventive and protective measures and monitor and review – **a suitable and sufficient risk assessment**.

A risk assessment that is 'suitable and sufficient' is defined in the ACOP as follows:

- It should identify the significant risks arising out of work.
- It should enable the employer or self-employed person to identify and prioritise the measures that need to be taken to comply with the relevant statutory provisions.
- It should be appropriate to the nature of the work and such that it remains valid for a reasonable period of time.

> ✎ *We will be looking at risk assessment in more detail later.*

The duty to assess risks under these Regulations is a general duty to cover all eventualities arising at and from work. If a more specific Regulation – e.g. DSE – applies to an activity or situation, it will not be necessary to repeat the existing risk assessment carried out for those Regulations, providing:

- **the assessment is still valid**
- **the assessment is 'suitable and sufficient'.**

> ✎ *Risk assessments are also required under the Fire Precautions (Workplace) Regulations 1997 (1999 amended).*
>
> - *Before employing young persons a full risk assessment must be carried out; competency and lack of training must be taken into consideration. Information, instruction and training must be built into a formal induction plan.*

Further requirements of these Regulations include the following:

- To make arrangements for putting into practice the health and safety measures identified by risk assessments, i.e. planning, organisation, control, monitoring and review.
- To provide health surveillance where necessary.
- To appoint **competent** persons to assist in undertaking the measures necessary to comply with statutory requirements.

> *A competent person is defined as someone who has sufficient training, experience, knowledge and understanding of the work involved, the capacity to apply this to the task required, an understanding of relevant current best practice and, most importantly, knows his/her limitations.*

- To set up emergency procedures for serious and imminent danger and danger areas.

> *For most employers fire and possibly bomb risks will be the main areas that need to be covered. However, other hazardous work activities may pose additional risks.*
> *Identify competent people to assist, compile written procedures and communicate to all staff – practise, practise, practise!*

- To ensure that necessary local contacts with external agencies are arranged.
- To provide employees with comprehensive and relevant information on health and safety matters.
- To co-operate with any other employer who shares a work site.
- To provide health and safety information to people working in their undertaking who are not their employees (the provision of risk assessments for contractors etc, and the obtaining of risk assessments from contractors).

> *It is important to have an exchange of documents before work commences. You should also have a list of approved contractors who are aware of your organisational requirements and can satisfy them.*

- To ensure that employees are adequately trained in health and safety matters and that they are capable of carrying out their work duties without risk to themselves or others.

> *It is the duty of employers to appoint capable, competent staff – the right person for the right job!*
> *Employers should not employ a person on a particular task for which he/she is not qualified or is insufficiently experienced, where this would possibly endanger his/her fellow employees (Davie* v *New Merton Board Mills* [1959]*).*

- Temporary workers and employment business workers must be given adequate information as regards health and safety.

> 💣 *You have the same duty to temporary staff, young people on work experience and volunteers regardless of how long they are going to be with you.*
> ✎ *Effective induction is the key issue here! Don't forget those who have been at your workplace for some time; they also need to know.*

- Consultation must take place with safety representatives.

> ✎ *'A Guide to the Health and Safety (Consultation with Employees) Regulations 1996' – this is covered in more detail later in this chapter.*

- Requirements to manage the health and safety of new and expectant mothers at work also have to be made. This affects women of childbearing capacity (not our definition!) where the work could, by reason of her condition, risk the health and safety of a new or expectant mother or her baby.

> ✎ *Systems should be established so that the employer is given the earliest possible notification of pregnancy, or when a mother has recently given birth and is in the workplace or is breast feeding, in order to enable: a) legal compliance, and b) prevent any harm to the woman or her unborn baby. If it is established that a new or expectant mother is at risk she must be moved to suitable alternative work, or if no such work exists, be suspended on maternity grounds (full pay and benefits).* ***Risk assessment is needed!***

- Protection of young persons – must take into account lack of experience and training to carry out the activity

General duties of employees

The Regulations complement the HASWA employee's duties with the following requirements:

- Use correctly all work items provided by the employer in accordance with training and instruction.
- Co-operate with the employer.

- Notify the employer of any work situation which might present a serious and/or imminent danger.
- Inform the employer of any shortcomings in the health and safety arrangements, for example where the risk assessment is inadequate.

Management Regulations in brief:

• Risk assessment and recording	reg. 3
• Health and safety arrangements	reg. 5
• Employers to arrange for health surveillance	reg. 6
• Health and Safety assistance	reg. 7
• Procedures for serious danger	reg. 8
• Contacts with external services	reg. 9
• Information for employees	reg. 10
• Co-operating co-ordination	reg. 11
• Non-employees on site	reg. 12
• Capabilities and training	reg. 13
• Employee's duties	reg. 14
• Temporary workers	reg. 15
• Expectant mothers	reg. 16/17/18
• Protection of young persons	reg. 19

✎ *These regulations are being constantly updated as and when new legislation becomes effective.*

Workplace (Health, Safety and Welfare) Regulations 1992

These Regulations replace a number of existing pieces of law, including sections of the Factories Act 1961 and the Offices, Shops and Railway Premises Act 1963. They cover most workplaces, including educational establishments, but not means of transport, construction sites, mines, quarries and fishing boats.

The Regulations are split into four broad areas:

Working environment

Temperature, ventilation, room space, lighting, etc.

Safety

Safe passage of pedestrians and vehicles, windows (cleaning, safe opening and closing), floors, doors/gates, falls and falling objects.

Facilities

Sanitary accommodation, clothing storage and rest areas.

> ✎ *Rest facilities must include arrangements to protect non-smokers from tobacco smoke. Also, rest facilities should be provided for pregnant women and nursing mothers.*

Housekeeping

Maintenance of the workplace, equipment and facilities, cleanliness and removal of waste materials.

> ✎ *It is interesting to note that these Regulations apply generally to employers only; there are no specific duties for employees. However, the 'catch all' approach would come from HASWA (s. 7) and the Management Regulations 1999.*

Workplace Regulations in brief:

- Workplace maintenance reg. 5
- Ventilation reg. 6

> ✎ *You need to provide enough fresh, breathable air.*

- Temperature reg. 7

> ✎ *This should be reasonable, with a readily available means of identifying the temperature; there is no stated maximum!*

- Lighting reg. 8

> ✎ *This should be suitable and sufficient for the purpose.*

- Cleanliness and waste materials reg. 9

> ✎ *Waste debris must be removed daily.*

- Space reg. 10
- Workstations reg. 11
- Conditions of floors and traffic routes reg. 12
- Falls and falling objects reg. 13

> ✎ *You must prevent falls of 2m or less if risk exists.*

- Windows regs. 14/16
- Traffic routes reg. 17

> ✎ *Pedestrians/vehicles to be separated by some means.*

- Doors and gates reg. 18
- Sanitary conveniences reg. 20
- Washing facilities reg. 21
- Drinking water reg. 22

> ✎ *You should provide a good wholesome supply of water.*

- Accommodation for clothing/changing regs. 23/24
- Rest and eating facilities reg. 25

> ✎ *It is well worth purchasing a copy of the ACOP relating to these Regulations, as they provide an in-depth look at the requirements relating to the workplace environment.*
> ✎ *Leaflets relating to the workplace can be obtained from the HSE.*

Provision and Use of Work Equipment Regulations 1992 amended 1998 – (PUWER)

The HSE guidance to these Regulations states that virtually all the requirements of PUWER already exist somewhere in law or constitute good practice. The Regulations bring together these requirements and apply them across all industrial, commercial and service sectors. This means that employers with well-chosen and well-maintained equipment should need to do little more than before.

The Regulations, which came fully into force on 1 January 1997, place on the employer the prime responsibility for the health and safety of workers using work equipment, and set a number of general duties such as:

- suitability of work equipment

> 💣 *Work equipment includes any machinery, appliance, apparatus or tool and any assembly of components which, in order to achieve a common end, are arranged and controlled so that they function as a whole.*
> ✎ *Lifting equipment is also subject to the requirements of Lifting Operations and Lifting Equipment Regulations 1998 (LOLER).*

- conditions in which it will be used

> 💣 *These must be suitable for the task.*

- maintained in good order

> ✎ *Maintenance logs to be kept – how else can you prove you have maintained it?*

- inspected to ensure that it is, and continues to be, safe for use

> ✎ *Inspections to be carried out by competent persons and records of inspection including dates etc to be kept.*

- the provision of adequate information, instruction and training to users

> 💣 *And their supervisors!*
> ✎ *It is a good idea to keep training records for many reasons besides being able to provide evidence to any enforcement officer – 'personal training logs'.*

- the provision of equipment that conforms with EU product Directives.

There are a number of specific requirements within these Regulations covering:

- guarding

> 💣 *Follow the hierarchy of guarding method – 'FIAT':*
> ***F**ixed closed/distance*
> ***I**nterlocking*
> ***A**utomatic/Adjustable*
> ***T**rip*
> 💣 *'Reasonably practicable' (cost v risk) in this instance does not apply – you have to take best option.*

- protection against specific hazards
- work equipment parts and substances at high or very low temperatures
- control systems
- isolation from sources of energy
- maintenance operations
- warnings and markings.

Hazards associated with machinery guarding can be classified into two main groups – mechanical and non-mechanical – and they all need to be considered when assessing work equipment:

Mechanical hazards	**Non-mechanical hazards**
entanglement	*electricity*
friction	*dust*
cutting	*noise*
stabbing	*hydraulics*
crushing	*vibration*
shearing	*fumes*
drawing in	*vapours*
impact	*gases*
abrasions	*lubricating fluids*

PUWER in brief:

- Suitability of work equipment — reg. 4
- Maintenance — reg. 5
- Inspection — reg. 6

- Specific risks reg. 7
- Information and instruction reg. 8
- Training reg. 9
- Dangerous parts of machinery reg. 11
- Protection against specific hazards reg. 12
- High/low temperature reg. 13
- Controls and control system reg. 14/18
- Isolation reg. 19
- Stability (fixed/mobile) reg. 20
- Lighting reg. 21
- Maintenance operations reg. 22
- Markings reg. 23
- Warnings reg. 24

> *There are no specific employee's duties. As with the Workplace Regulations the 'catch all' comes under the HASW etc Act 1974 and Management Regulations 1999.*

Manual Handling Operations Regulations 1992

Manual handling means more than just lifting; it is defined as the transporting or supporting of a load which includes lifting, putting down, pushing, pulling, carrying, moving by hand or bodily force.

Research from the Health and Safety Executive shows that 34% of all accidents reported each year involve manual handling. This equates to approximately £100m in lost production and 33 million working days lost in the UK.

> *Manual handling accidents are a feature in all work activities regardless of sector. For example, 30% of all accidents suffered in the banking and finance sector are due to manual handling.*

General duties of employers

These Regulations apply to all manual handling tasks involving a risk of injury.

Hazardous manual handling operations should be avoided wherever possible. For example, could a load be moved by other means?

Where the manual handling of loads cannot be avoided, employers must organise workstations (Workplace Regulations) where manual handling can take place as safely as possible. Employers must risk assess the health and safety conditions of the type of work involved and take appropriate measures to avoid or reduce the risk. In doing this they must take account of the following:

- requirements of the task

> ✎ *Height; distance; obstacles; route taken.*

- characteristics of individuals

> ✎ *Personal capabilities; height; weight; special circumstances; clothing required; knowledge and training.*

- characteristics of the load

> ✎ *Shape; size; weight; temperature; centre of gravity.*

- environment

> ✎ *Lighting; space involved; doors; temperature; obstructions.*

> ✎ *The mnemonic **TILE** is a useful device for remembering the above:*
> ***T**ask*
> ***I**ndividual*
> ***L**oad*
> ***E**nvironment*
>
> 💣 *Avoid manual handling wherever possible or eliminate by other means, then use **TILE** – the risk of injury should be reduced to the lowest level that is reasonably practicable.*
> ✎ *Look out for the abnormal activity when manual handling – for example, moving office furniture where desks, computers, etc are carried – which can often lead to injury. In these sorts of abnormal situations it is useful to have some trained members of staff who can assess the activity and give appropriate advice.*

Any assessment made shall be reviewed by the employer if there is reason to suspect that it is no longer valid or there has been a significant change in working practices.

General duties of employees

Employees have specific duties to make full and proper use of any equipment or system provided for manual handling.

> ✎ *These duties are reinforced by those general duties placed on employees by s. 7 of the HASW etc Act 1974 and also reg. 14 of the Management Regulations 1999*

Manual Handling Regulations in brief:

- Employer's duties reg. 4
- Employee's duties reg. 5

> ✎ *For best practice, train all staff in the following lifting techniques:*
> - *proper balanced positioning of feet*
> - *arms in/chin in*
> - *keep load close to body*
> - *walk in direction of movement*
> - *put load down and then adjust.*

Personal Protective Equipment at Work Regulations 1992 (PPE)

Personal protective equipment (PPE) includes both of the following, when they are worn for protection of health and safety at work:

- **Protective clothing** – aprons, gloves, safety footwear, safety helmets, high-visibility waistcoats, protective clothing for adverse weather conditions.
- **Protective equipment** – eye protectors, life jackets, respirators, underwater breathing apparatus and safety harnesses.

The Regulations will not apply to ear protectors or some other types of PPE used at work, as they are covered by more specific legislation: e.g. Noise at Work Regulations 1989. However, the advice given in these Regulations may still be applicable in relation to selecting and maintaining PPE and training employees in usage.

> *PPE should always be regarded as 'last resort' – engineering controls and safe systems of work must be considered first.*
>
> *It is the employer's duty to provide a safe place first, then a safe person.*

Also excluded are such things as ordinary clothing, PPE for road transport and sports equipment, etc.

General duties of employers

Employers have the following duties:

- To carry out risk assessment and organise the selection of suitable PPE – risk assessment to be designed to cover those risks which have not been prevented by other means.
- To review the risk assessment when they suspect that it is no longer valid or when significant changes have affected the matters to which it relates.
- To provide suitable PPE.
- To ensure the compatibility of PPE. If more than one item of PPE has to be worn then the PPE should be comfortable, compatible and continue to provide effective protection against risk.
- To make sure that PPE is effectively maintained.
- To provide appropriate accommodation for PPE when it is not in use.
- To provide adequate information, instruction and training.
- To ensure adequate supervision.
- To take steps to ensure that PPE is properly used.

> *No charge can be made on the individual for the provision of PPE (s. 9 of the HASW etc Act 1974).*
>
> *Employers must assess risk and not rely on PPE if other measures can be taken.*

General duties of employees

It is the duty of employees to use any PPE provided in accordance with instructions and training, and those who have been provided with PPE shall report any loss or obvious defect immediately.

> 💣 *As a designated leader you should always ensure that any necessary PPE is worn as required by the risk assessment.*
>
> 💣 *You are taking a grave responsibility upon yourself if you allow employees to disregard the PPE requirements. The tightening up of supervision in this area may become 'uncomfortable' for some organisations that have allowed discipline to become slack.*
>
> ✎ *Take the role of responsibility seriously, and discipline as appropriate.*

PPE in brief:

• Provision – last resort!	reg. 4
• Multi-PPE compatibility	reg. 5
• Assessment of PPE	reg. 6
• Maintenance/replacement	reg. 7
• Storage facilities	reg. 8
• Information, instruction, training	reg. 9
• Use of PPE	reg. 10
• Reporting loss/defect	reg. 11

Health and Safety (Display Screen Equipment) Regulations 1992 (DSE)

> ✎ *You are strongly advised to obtain from the HSE their booklets relating specifically to these Regulations.*

The Regulations apply to all workstations with a display screen mainly used for text, line drawings and graphics, and to personnel who use them as a significant part of the job, although there are certain exceptions:

- drivers' cabs or control cab systems
- on-board computer systems

- portable systems not in prolonged use at a workstation
- cash registers

> *Under the Health and Safety Amendment Regulations 2002 all workstations are now covered – there is no distinction between 'operator' and 'user'.*

General duties of employers

The requirements of these Regulations include:

- analysis of workstations and identification of measures to reduce risks to lowest extent reasonably practicable
- requirements for workstations
- daily routine of users
- eye and eyesight test
- provision of information and training.

> ✎ *If you employ temps the responsibility for eye testing etc lies with the agency; the reponsibility for providing information and training etc lies with yourselves (Health and Safety Misc Amendment 2002).*
>
> ✎ *When assessing workstations employers must take into account the **individual**; **equipment** (display screen, keyboard, work desk, work chair); **environment** (space, lighting, noise, heat and humidity); **interface** (suitability, adaptability, format and pace) and **training** (trained in safe use of, informed of risks, postural attitude).*
>
> ✎ *Employers must consider the employee who works at home and also laptop users in their risk assessment.*

General duties of employees

There are no specific requirements in relation to employees under these Regulations. However, other legislation does apply, such as s. 7 of the HASW etc Act and reg. 14 of Management Regulations 1999. It is essential to point out under this heading that display screen equipment is not thought to cause any risk to health or safety. The risk usually occurs through misuse due to a lack of understanding, training and relevant information – **a management issue!**

Problems usually associated with working with display screen equipment are eye strain, constant headaches, repetitive strain injury, work-related upper-limb disorders and, perhaps to a lesser degree, back problems, as well as general fatigue and stress.

✎ *As managers you need to educate and manage your staff so that they can avoid these potential problems.*

✎ *Train users to risk assess their own areas, because they need to understand the importance of setting up their workstation and work routine in the context of the long-term effects of health and safety.*

💣 *When assessing workstations you must bear in mind previous legislation: e.g. Workplace Regulations (lighting, temperature, housekeeping) and Management Regulations (information, training and employee's responsibilities).*

💣 *Don't forget to assess laptops, look at providing 'docking stations' and guidance notes on how to set up correctly, etc.*

DSE in brief:

- Scope of user/equipment — reg. 1
- Analysis of workstations — reg. 2
- Workstation requirements — reg. 3
- Daily work routine of users — reg. 4
- Eye and eyesight — reg. 5
- Training — reg. 6
- Information — reg. 7

Control of Substances Hazardous to Health Regulations 2002 (COSHH)

A substance can be defined as any form/type that can present a hazard to health that is used or generated out of, or in connection with, any work activity under the employer's control. It can take the form of fumes, gas, chemicals, vapours, dust, micro-organisms, bacteria, yeast and moulds. There may be others!

✎ *What we need to assess is how substances can harm people. Therefore we must consider the route of entry into the body – typically absorption (through the skin), injection, inhalation and ingestion – and the effects of these substances on the organs within.*

These Regulations are designed to protect people against risks to their health, whether immediate/short term (acute) or delayed/long term (chronic), from substances hazardous to health arising from a work activity. The substances referred to are those listed in the Chemicals (Hazard Information and Packaging for Supply) Regulations 1994 (CHIP2) (updated CHIP3 2002) which are classified as being very toxic, toxic, harmful, corrosive, sensitising or irritant, and also those that have a maximum exposure limit (MEL) or occupational exposure standard (OES).

✎ *A list of MELs and OESs can be found in HSE document EH40 Occupational Exposure Limits.*
💣 *MEL – the maximum exposure limit of an airborne substance that an employee may not be exposed to by inhalation under any circumstances.*
💣 *OES – the concentration exposure limit of an airborne substance at which there is no evidence that it is likely to be injurious by inhalation.*

✎ *It is essential that information is kept up to date on all chemicals used within the workplace, as scientific evidence on the hazards to health and the associated risks is constantly changing.*

For further reference the HSE have produced 'COSHH Essentials (HSG 193) – easy steps to control chemicals' www.coshh-essentials.org.uk.

General duties of employers

- Prohibition of certain harmful substances.
- An employer may not carry on any work which is liable to expose any employee or other person to a substance hazardous to health, unless a suitable and sufficient assessment has been made of the risks to health created by that work and the necessary measures to control exposure.

💣 *Risk assess*
⇨ *what hazards exist*
⇨ *what form these hazards take*

> ⇨ *who will be affected*
> ⇨ *how often and how seriously*
> ⇨ *which workplace*
> ⇨ *what monitoring procedure exists*
> ⇨ *what do employees/others need to know.*

- Every employer shall ensure that the exposure of employees to hazardous substances is either prevented or adequately controlled.

> *Hierarchy of control*
> ⇨ *avoid by other means*
> ⇨ *substitute for other less hazardous substances*
> ⇨ *eliminate time of exposure/people/small quantities*
> ⇨ *isolate*
> ⇨ *ventilate (local exhaust ventilators)*
> ⇨ *control – safe working methods, safe handling, storage*
> ⇨ *PPE*
> ⇨ *designated areas, hygiene measures.*
> *A management system for discipline is essential within these Regulations – substances can kill!*

- Any control measure provided (e.g. local exhaust ventilators) should be maintained in efficient working order and in good repair (at lest once every 14 months).

> ✎ *You need to keep maintenance records (PUWER).*

- Routine monitoring should be carried out. Records must be kept showing when monitoring was done, the findings and what measures were put in place.
- Health surveillance must be carried out if the employee is exposed to any of the substances listed within the Regulations, and records of the surveillance kept for at least 40 years (if the employee is identifiable) or five years in any other case, from the date of the last entry.

> *If the organisation ceases trading then the records must be offered to the HSE.*

- Pay for health surveillance checks.
- Provide relevant information, instruction and training to persons exposed to hazardous substances within the working environment.

✎ *For every substance in use in the workplace a hazard data sheet should be provided which gives information on:*

- *what the substance is*
- *the precautions to be taken when using it*
- *emergency procedures in the event of a spillage*
- *first-aid procedures in the event of an accident involving a substance.*

These hazard data sheets must be kept up to date for all substances in use within the workplace and be available to those users.

✎ *Those with the responsibility should ask themselves the following questions:*

- *what is the risk?*
- *can it be eliminated?*
- *can we substitute it for something less hazardous?*
- *can we contain it?*
- *can we protect people from it?*
- *what are the secondary risks (fire, explosion)?*
- *should we therefore be using it?*

✎ *Once the organisation has assessed all its substances, provided all the controls necessary including hazard data sheets and all relevant information for employees, the list can be regarded as 'approved'. The organisation will therefore benefit form having a strict buying policy, so that no unapproved substance can find its way into the workplace. Any new substance required should go through the COSHH assessment procedure.*

- Have arrangements in place to deal with accidents, incidents and emergencies.

General duties of employees

- To co-operate at all times.
- To attend health surveillance checks during working hours if required.
- To have access to health surveillance records.
- To wear suitable PPE as appropriate.
- To make full and proper use of any control measures provided.
- To report any defects in equipment provided.
- To store correctly in accommodation provided.

COSHH in brief:

• Duties of employer to others	reg. 3
• Prohibited substances	reg. 4
• Assessment	reg. 6
• Prevention or control	reg. 7
• Use of control measures	reg. 8
• Maintenance of control measures	reg. 9
• Monitoring exposure	reg. 10
• Health surveillance	reg. 11
• Information, instruction, training	reg. 12
• Arrangements to deal with accidents, incidents and emergencies	reg. 13
• Fumigation	reg. 14

Electricity at Work Regulations 1989

Electricity, properly used, is a safe, convenient and efficient source of energy. Misused, or allowed to get out of control, it can cause damage, injury or death.

The Electricity at Work Regulations 1989 apply to all workplaces, including educational establishments and laboratories, and cover all electrical systems and equipment. The main purpose of the Regulations, which came into force on 1 April 1990, is to ensure that employers and employees take precautions against the risk of death or personal injury from electricity at work. The Regulations are directed at users of electricity rather than suppliers or manufacturers of electrical equipment.

The Regulations define 'danger' as the risk of injury and go on to say that 'injury' means death or injury caused by *electric shock, electric burn, fires of electrical origin, electric arcing, explosions initiated by electricity or other secondary effects.*

> *It is important to note that within the Regulations varying levels of duty are specified:* ***absolute; practicable; reasonably practicable.***

General duties of employers

- All electrical equipment shall at all times be of such construction so as to prevent danger.
- All electrical equipment and associated work activities to be maintained so as to prevent danger.

> *This includes portable equipment (discussed later in this chapter).*

- Any equipment provided for the protection of persons working on or near electrical equipment to be maintained.

> *Regular inspection of all electrical equipment is an essential part of any preventive maintenance programme. Equipment includes wiring installations, generators or battery sets and everything connected to them.*
> *A planned electrical maintenance system will include a procedure for the periodic inspection, testing and repair to be carried out before it becomes defective – choosing the right interval is crucial.*
> *Records of maintenance and test results of every piece of electrical equipment will enable effective monitoring.*

- All equipment to be protected so as to prevent any danger to users when exposed to adverse effects (weather, temperature, pressure, dust/dirt, natural hazards).
- Ensure the correct insulation, protection and placing of conductors to prevent danger.
- Ensure earthing or other suitable precautions to prevent danger resulting from any conductor becoming charged (this may also include the conductive parts of equipment such as outer metallic casings which, although not live, may become live under fault conditions).

> *Techniques employed to achieve the above may include double insulation, use of safe voltages (reduced voltage systems), earth-free non-conducting environments, separated or isolated systems.*

- Joints and connections shall be mechanically and electrically suitable for use.

> *Special attention should be given to joints and connections in cables, plugs and sockets (portable equipment).*

- Install means of protecting from excess of current within every part of a system, usually in the form of fuses or circuit breakers.
- Have in place readily available means of cutting off and isolating the supply.

> *Switching off can be by direct manual operation or by indirect operation via stop buttons. Isolation should be achieved by ensuring that the supply will remain switched off and reconnection prevented.*

- Precautions to be taken when working on equipment made 'dead'.

> - *Permit to work! (see Chapter 5).*
> - *In some instances it is necessary to work on 'live' equipment, such as fault diagnosing and equipment testing – in these and other similar cases it is imperative that safe systems of work are followed and levels of competencies clarified.*

- No person to be allowed to work on or near any live conductor not suitably covered with insulating material (this includes underground and overhead cables).

> *Permit to work. Use of competent persons only!*

- Adequate working space, means of access and lighting when working with or near electrical equipment, where work is being carried out in circumstances that may give rise to danger.

> *Space – allow enough room for persons to pass freely and be able to pull away from danger.*

💣 *Lighting – this should be natural if possible. Any artificial lighting to be permanent and a properly designed installation. Where not possible, then the level of lighting must be adequate to prevent injury.*

- Persons employed on any work activity where technical knowledge or experience is necessary to prevent danger or injury must be competent or suitably supervised.

💣 *Competency (technical knowledge or experience) may include:*

- *adequate knowledge and experience of electricity*
- *an understanding and practical experience of the system to be worked on, hazards which may arise during the work and precautions which need to be taken*
- *ability to recognise at all times whether it is safe for work to continue.*

✎ *Conduct a suitable and sufficient risk assessment.*
💣 *Supervision is essential.*
✎ *The responsibilities of supervision should be verbally communicated if 'low risk' and consideration should be given to communicating responsibility in high-risk or complex working areas in writing.*
💣 *Over half the people involved in electrical accidents were not competent to undertake the task (source:* Electrical Incidents in Great Britain Statistical Summary*).*

General duties of employees

- To co-operate with the employer in meeting the requirements of the Regulations.
- To comply with the Regulations.

✎ *Employees also have duties under ss. 7/8 of the HASW etc Act 1974 and Management Regulations 1999 (reg. 14).*

Portable equipment

Generally, portable electrical equipment describes appliances that have a lead (cable) and a plug and can easily be moved around.

About a quarter of all reported electrical accidents involve portable equipment. Typical causation factors include faulty or frayed cables, loose leads, defective plugs and sockets. Also, the use of unsuitable or defective equipment, inadequate maintenance and lack of training add to the accident toll.

Portable equipment maintenance should involve visual inspection, testing, repair and replacement. A cost-effective maintenance programme can be achieved by a combination of the following:

- visual checks by the user
- visual inspections by an appointed person
- combined inspection and tests by a competent person.

Checklist

Visual

- Check damage to cable (cuts, fraying cables, cable grips, signs of overheating).
- Check damage to plug/extension leads (bent pins, broken casing).
- Check damage to equipment (loose connections, screws missing, burn evidence).
- Assess the suitability of equipment under conditions being used.

Formal

- As above (formally).
- Test equipment (polarity, fusing).
- Assess the suitability of equipment within the environment.

(The 'formal' inspection/testing should be carried out by someone with a wider understanding/competence than the person carrying out visual inspections or the less formal visual inspection carried out by the user.)

✎ *The frequency of maintenance/testing of any portable equipment will depend on the risk assessment. Factors to be taken into account should include:*

- *type of equipment, how it is used and the environment in which it operates*
- *manufacturer's recommendations*
- *age and reliability of the equipment.*

(Guidance note '*Maintaining Portable and Transportable Electrical Equipment*' (IND(G) 236L) can be purchased form the HSE.)

Electricity Regulations in brief:

- All systems/work activities prevent danger/ properly used — reg. 4
- Strength and capabilities — reg. 5
- Adverse/hazardous environment — reg. 6
- Insulation/conductors — reg. 7
- Earthing or other precautions — reg. 8
- Integrity of circuitry — reg. 9
- Joints/connections — reg. 10
- Protection from excess current — reg. 11
- Cutting off/isolation of supply — reg. 12
- Precautions – equipment made dead — reg. 13
- Working on or near live conductors — reg. 14
- Working space/access/lighting — reg. 15
- Competent person/supervision — reg. 16

Noise at Work Regulations 1989

Noise is one of the most common and potentially damaging health problems faced within industry.

Noise is often described as unwanted sound or a subjective response to sound. The effects of 'noise' to health can range from a **temporary threshold shift**, resulting from exposure to fairly high noise levels for short periods (experienced following a loud rock concert!), to **permanent threshold shift**, caused by prolonged exposure to loud sounds which eventually damage the inner ear. There is no recovery from permanent threshold shift. Another noise-induced illness is tinnitus – ringing noises in the ear. A shift in the threshold can also be caused by age.

Apart from the physical effects, noise can also be quite irritating, causing lapses in concentration, errors in work, stress-related illnesses, masking of important warning signals (fire alarms, warning of vehicle approach), all of which contribute to the number of accidents in the workplace.

The Noise at Work Regulations 1989 gave the HSE a firm base for enforcing industry to reduce the number of employed persons exposed to noise following the official recognition of noise-induced hearing loss (occupational deafness) as an accepted industrial disease (National Insurance [Industrial Injuries] Act 1975). Other statutory requirements relating to noise at work can be found within the HASW etc Act 1974 – *the provision and maintenance of a working environment that is, so far as is reasonably practicable, safe and without risks to health.* Employers also have an additional pressure under the Management Regulations 1999 – *every employer shall ensure that his employees are provided with such health surveillance as is appropriate, having regard to the risks to their health and safety which are identified by the assessment.*

General duties of employers

- To make a formal noise assessment where employees are likely to be exposed to the first action level (85dB(A)) or above, or to the peak action level (200 pascals) or above. The assessment should be carried out by a competent person.
- To identify which employees are exposed and provide information.
- To review any assessments when significant changes in the work have taken place or when the original assessment is no longer valid.
- To keep records of all assessments made.
- To reduce the risk of exposure to the lowest level reasonably practicable.
- To reduce, so far as is reasonably practicable, the exposure of employees to second action level (90dB(A)) or above, or peak action level or above, by means other than the provision of personal ear protection.

💣 *To create a 'safe place'*
⇨ *eliminate noise*
⇨ *reduce at source by engineering means*
⇨ *isolate noise or user*
⇨ *control: time, frequency of exposure.*

To create a 'safe person'

💣 *PPE*
⇨ *discipline!*
✎ *This is a management issue!*

- To provide ear protection when an employee is exposed between first action level and second action level, at the employee's request.
- To provide suitable ear protection when an employee is exposed between second action level and peak action level or above which will bring the level down.
- To identify ear protection zones with mandatory signs where areas exist between second action level and peak action level or above.
- To ensure all employees and others entering any such above area wear personal ear protectors.
- To ensure correct use of equipment and that it is fully maintained.
- To provide adequate information, instruction and training to every employee exposed between first action level and peak action level or above on the risks of hearing damage; precautions to take; how to obtain ear protectors; obligations of the employee.

General duties of employees

- To make full and proper use of personal ear protectors when provided.
- To make full and proper use of any other protective measures provided.
- To report any defects to the employer.

Noise control

There are three basic ways of limiting or controlling noise, which may affect the health of employees and others at work:

- reduction of noise at source
- isolation of the source/transmission path
- protection of the receiver.

The source

In engineering terms, it is feasible to produce quieter machines, but it is also about reducing the strength of the source. For example:

- **The movement of air** – providing silencers, reducing air pressures and smoothing air flows.
- **The force of impact** – produced in hammering, riveting or with rattling components, which can be treated by damping, cushioning, lowering impact mass and a good maintenance regime.
- **Working forces** – produced in cutting processes etc, which can be treated by reducing the speed of equipment, taking finer cuts, damping down the process.
- **Cyclic** – caused by motor vibration, gear wear, ineffective pumps and bearings, which when out of balance or poorly maintained can cause excessive noise.

The transmission path

Another point to consider when controlling noise is the transmission path. Can the noise be reflected to reduce its strength, insulated by materials or distance from source to receiver, or absorbed?

The receiver

What about the receiver? Is the use of PPE a failure in management responsibility – the last resort? Should automation be considered first, followed by isolation of the equipment or receiver (noise havens, silenced control areas) or, as part of the management process, should we be looking at reducing the time and doubling the distance of exposure to the noise before PPE is considered?

Noise at Work Regulations in brief:

- Definitions reg. 2
- Assessment reg. 4

• Records of assessment	reg. 5
• Reduction of risk	reg. 6
• Reduction of exposure	reg. 7
• Ear protection	reg. 8
• Ear protection zones	reg. 9
• Maintenance and use	reg. 10
• Information, instruction, training	reg. 11

It is important to note here that the action levels may be changing (by 2006). First action level reduced to 80 dB(A); second action level to 85 dB(A) and peak action being reduced to 140 pascals and a new weekly exposure level limit value of 87 dB(A) (visit HSE website for updates).

Fire Precautions Act 1971 & Fire Precautions (Workplace) Regulations 1997/1999

Fire in Britain causes the death of approximately a thousand people every year. Many of these deaths occur in the workplace and are preventable.

Causes of fire range from the direct negligence of employees, poor maintenance, careless hot work, spillage or mixed storage of flammable liquids, to faulty equipment. Bearing this in mind the main aim of the Regulations is to make sure every employee knows what actions need to be taken in the event of fire and how to evacuate the premises quickly and safely.

The three essential points to be covered in instructions, drills and in the event of an outbreak of fire are:

- raising the alarm and calling the fire brigade
- immediate attack/action
- evacuation.

Fire Precautions Act 1971

The main requirement of the Fire Precautions Act 1971 was for designated premises to be 'fire-certificated', normally by the local fire authority, though in exceptional hazardous circumstances by the HSE. Designated premises are defined as the following:

- those providing sleeping accommodation
- those providing treatment of care
- places of entertainment, recreation
- those providing teaching, training or research
- those providing public access
- those providing a place of work.

Within the Act, places of work requiring a fire certificate are defined as:

- employing more than 20 people
- employing more than ten persons above/below the ground floor
- buildings of multiple occupation where the aggregate of people at work exceeds the above totals
- premises where explosives or highly flammable materials are stored or used in or under the buildings.

The fire certificate contains information concerning the use of the premises and also specifies the following:

- means of escape (as per the route specified on the drawings – usually attached)
- fire-fighting equipment
- fire-alarm system
- particulars concerning any explosives etc to be used or stored on the premises.

In addition, the certificate may require that fire-escape routes are kept clear, fire-fighting equipment and devices are maintained, people are trained in fire procedures and records kept.

General duties of employers

- To provide a means of escape.
- To provide a means of fighting the fire.
- To ensure that a fire certificate is kept on the premises (preferably displayed).
- To provide adequate training and keep records accordingly.
- To inform the fire authority of any changes to the building, internally and externally.

> *Additional duties under ss. 2/3 of the HASW etc Act 1974 and also Management Regulations 1999 – risk assessment of the workplace.*
> ✎ *A separate fire risk assessment, besides constituting good management practice, is a general requirement within the Fire Precautions (Workplace) Regulations 1997/1999 for those who do not already hold a fire certificate.*

To formally assess the risk of fire within any premises it is an advantage to understand in more detail certain aspects of precisely what is required:

- **Safe means of escape** – a structural means that forms an integral part of the structure and provides a safe escape route for persons using their own unaided effect to reach a place of safety.
- **Distance of travel** – the actual distance a person must travel between any point in a building and the nearest door to a protected route of final exit.
- **Protected route** – a route leading to an exit from a floor or to a final exit which is separated from the remainder of the building by walls, partitions, doors, floors and/or ceilings of fire-resistant construction.
- **Final exit** – the termination of an escape route from a building giving direct access to a place of safety and sited so that persons can disperse safely.

A designated travel route inspection should check for the following:

- that the route is well signed – clear, visible
- that it passes through a low-risk area
- that all passageways are unobstructed
- that routes are of sufficient width for the flow of people
- design of inner rooms – vision panels required
- that there are fire-resistant dead ends
- that there are smoke doors
- that emergency lighting has been installed, if necessary
- that the final exit doors should open out and remain unlocked
- that safe assembly points exist away from the risk.

Fire Precautions Act in brief:

● Premises requiring certificate	s. 1
● Contents and location of certificate	s. 6
● Inspection and changes to premises	s. 8
● Minimum requirement for premises exempt	s. 9
● Powers of inspection	s. 19

Fire Precautions (Workplace) Regulations 1997/1999

The Fire Precautions (Workplace) Regulations 1997/1999, which came into force on 1 December 1997, are designed to 'safeguard the safety of employees in case of fire' and apply to all workplaces (except private dwellings). There are exceptions to these Regulations, the principal one being those premises holding a current fire certificate under the Fire Precautions Act 1971.

General duties of employers

- Fire risks within the workplace to be assessed.

> ✎ *This can be done individually or as part of a general health and safety risk assessment.*
>
> 💣 *Fire risk assessment should include:*
>
> – *identification of combustible materials, usage and storage, etc.*
> – *identification of ignition sources: e.g. work equipment*
> – *structure of building*
> – *employees at high risk*
> – *record of findings and measures taken to reduce the risk.*

- To provide emergency routes and exits.
- To provide means of fighting and detecting fire.
- To ensure that the maintenance and testing of equipment is carried out by a competent person in accordance with manufacturer's instructions.

Fire Regulations in brief:

- Assess the risk of fire
- Detection and warning of fire
- Access/egress

- Appropriate training, information
- Fire-fighting equipment and fire-fighting measures
- Maintenance of equipment
- Emergency routes/exits
- Enforced by local authorities
- Risk assessment (these Regulations add to the risk assessment requirements under reg. 3 of Management Regulations 1999).

Health and Safety (First Aid) Regulations 1981

The Health and Safety (First Aid) Regulations 1981 came into effect on 1 July 1982. Together with the ACOP, which was revised in 1990 and more recently in 1997, they are framed to allow employers considerable scope in deciding how to tailor first-aid provision within the workplace.

The most important questions which should be considered are the number and type of first aiders to be employed and whether a first-aid room should be provided. The main criteria that apply here are the number and location of persons employed, the nature of the hazards to which they are exposed and the accessibility of emergency facilities.

First aiders

The ACOP suggests that every establishment should have one **appointed person** who is authorised to take charge of the situation if there is a serious injury or illness (contact with local ambulance). The person will act in the absence of a trained **first aider**, or where a first aider is not required (fewer than 50 employees in low-risk areas).

In **low-risk** areas, e.g. offices, libraries etc., one first aider is required for between 50 and 200 employees. **Medium-risk** areas of work, e.g. light engineering, warehousing, food processing etc, require an appointed person when employing fewer than 20 and at least one first aider between 20 and 100 employees. Within areas of **high-risk**, e.g. heavy industry, chemical manufacture etc, the ACOP suggests one appointed person for fewer than five employees and at least one first aider for between five and 50 employees. In addition to this, in any high-risk area where there is a risk of poisoning for which special treatment is required, then at least one first aider trained in specific emergency action is required.

✎ *The guidance given within the ACOP outlines the minimum requirements needed. It would be advisable for any organisation to risk assess their area of activity and appoint a sufficient number of first aiders based on the risk – the higher the risk the more first aiders are required.*

First-aid room

A first-aid room should be provided if the workplace has a high-risk environment, but the proximity of accident and emergency facilities should also be taken into account.

✎ *Requirements to have a sizeable first aid room for holding/ removing a stretcher.*

Training of first aiders

A 'suitable person' is a first aider who has undergone training approved by the HSE. In recruiting potential first aiders it is advisable that employers select from employees who are suitable to undergo first-aid training and able to leave their work immediately should an emergency occur.

✎ *Bear in mind holiday cover and shift work when appointing first aiders.*

First-aid boxes

All establishments are required to have at least one first-aid box, made of suitable material to protect the contents from damp and dust. The boxes should be clearly identified as first-aid containers and be positioned in a clearly marked and readily accessible place. They should be adequately stocked and contain only those items that a first aider has been trained to use.

Records

First aiders should keep records of all the cases they treat, the name of the patient, date, time and circumstances of the accident and details of injury suffered and treatment given.

> ✎ *This is an important point, as the organisation's insurers will need this information.*

General duties of employers

- To provide such equipment and facilities for first aid to be carried out.
- To provide a number of 'suitable persons' to carry out first aid.
- To provide training for 'suitable persons'.
- To nominate 'appointed persons' to assist when 'suitable person' absent.
- To inform employees of arrangements, including location of equipment, facilities and nominated personnel.

General duties of employees

As in other areas of legislation there is no specific requirement, but again the 'catch all' situation applies through the HASW Act etc 1974 and Management Regulations 1999

First Aid Regulations in brief:

- Definition: reg. 2
  - preserve life
  - prevent deterioration
  - promote recovery
- Employer's responsibility reg. 3
- Provision of information reg. 4
- Self-employed responsibilities reg. 5

Reporting of Injuries, Diseases and Dangerous Occurrences Regulations 1995 (RIDDOR)

Reporting of accidents, diseases and dangerous occurrences (both fatal and non-fatal) by employers and other 'responsible persons' allows the HSE to identify and monitor accident trends and occupational diseases. This enables it to take action, where needed, to improve the existing situation and prevent ill health in the workplace.

> *It is important to ensure that all notifiable accidents are actually reported. There have been prosecutions of organisations and individual managers who have failed to report notifiable accidents.*

RIDDOR 1995 extends RIDDOR 1985 by streamlining and simplifying the legislation on reporting serious workplace accidents, incidents and ill health, and also replaces five previous pieces of law applying separate reporting systems.

The main areas of improvement are that acts of violence done to persons at work and acts of suicide on railways, etc are now included along with reporting injuries to members of the public and also updated lists of reportable injuries, dangerous occurrences and notifiable diseases.

Definitions

Injuries or conditions include:

- specified fractures (other than to fingers, thumbs or toes) and amputations
- dislocation of shoulder, hip, knee or spine
- loss of sight (temporary or permanent)
- injury caused by electrical equipment
- loss of consciousness due to lack of oxygen
- decompression sickness
- illness resulting from inhalation, ingestion or absorption of any substance
- illness thought to have been caused by exposure to a pathogen or infected material
- any injury causing the person to be admitted to hospital for more than 24 hours
- acts of non-consensual physical violence done to a person at work
- death or injury, where a member of the public is taken to hospital.

Reportable diseases include:

- certain poisonings
- skin diseases, occupational dermatitis, skin cancer, chrome ulcer
- lung diseases, occupational asthma, pneumoconiosis, mesothelioma
- various musculoskeletal disorders.

Dangerous occurrences include:

- collapse or failure of any load-bearing parts of a crane or access platform
- explosion or failure of a pressure vessel
- electrical short circuit attended by fire and plant stoppage for 24 hours
- explosion or fire in plant, with plant stoppage for 24 hours
- sudden uncontrolled release of highly flammable liquid or gas
- collapse of scaffold more than 5m high or of more than 5 tonnes of building.

General duties of employers

- To provide employers with clear guidelines on reporting procedures.
- To notify the enforcing authority by the quickest practicable means within ten days of accident/incident occurring.

> ✎ *The quickest possible means can be by telephone – keep a record of telephone notifications, including time of call, name of recipient and details given.*

- To fill in approved form – F2508.
- To report the death of an employee in writing to the enforcing authority as a result of an accident at work which caused the death within one year of the accident.
- To appoint a responsible person to report to the enforcing authority if a person suffers from any specified occupational disease.
- To report any gas (mains, LPG) incident causing death or injury – form F2508G.
- To keep records of every event reported for three years.
- To ensure that records are made available to the HSE and safety committees on request.

RIDDOR in brief:

- Person responsible for reporting — reg. 2
- Notification and reporting — reg. 3
- Reporting of death of employee — reg. 4

- Reporting disease reg. 5
- Reporting gas reg. 6
- Records reg. 7

The HSE has recently set up a Single Incident Contact Centre (ICC) as an alternative to enable reporting quicker. Users may telephone (0845 300 9923) fax reports (0845 3009924) or email: riddor@natbrit.com

Safety Representatives and Safety Committees Regulations 1977 & Health and Safety (Consultation with Employees) Regulations 1996

Consulting employees on health and safety matters is very important in creating and maintaining a safe and healthy working environment. By consulting employees, an employer should be able to motivate staff and make them aware of health and safety issues.

Consultation involves employers not only giving information to employees but also listening to and taking account of what employees say before they make any health and safety decisions.

If a decision involving work equipment, processes or organisation could affect the health and safety of employees, the employer must allow time to give the employees or their representatives information about what is proposed. The employer must also give the employees or their representatives the chance to express their views and take these into account before reaching a decision.

Safety Representatives and Safety Committees Regulations 1977

If an employer recognises a trade union and that trade union has appointed, or is about to appoint, safety representatives, then the employer must consult those safety representatives on matters affecting the group(s) of employees they represent.

Safety Representatives and Safety Committees Regulations in brief:

- Appointment of representative reg. 3
- Functions of representative reg. 4
- Authority to inspect reg. 5

- Inspection following accident reg. 6
- Inspection of documentation reg. 7
- Safety committees reg. 8
- Appeals reg. 11

Health and Safety (Consultation with Employees) Regulations 1996

Any employees not in groups covered by trade union safety representatives must be consulted by their employers either directly or through elected representatives.

Consultation with Employees Regulations in brief:

- Duty of employer to consult reg. 3
- Person to be consulted reg. 4
- Provide information reg. 5
- Elected representatives reg. 6
- Training of representatives reg. 7

Under both sets of Regulations the employer has a duty to make sure that the elected representatives receive the necessary training to carry out their role effectively, give them the necessary time off with pay and pay any costs incurred.

> ✎ *Briefing all employees about the requirements of health and safety is of paramount importance. Communication can be by written format – i.e. notice boards, posters, memos, letters – or face-to-face – i.e. verbal, meetings, briefing system, presentations. You need to decide which way is appropriate for the level of importance, perhaps using a variety of methods.*

CHAPTER 4

Risk assessment – the proactive basis for managing health and safety at work

Every organisation must carry out health and safety risk assessments: there is no choice, it is a clear legal requirement.

Probably the most important factor as to how successful your risk assessment programme will be is the extent that you involve those who actually carry out the tasks within your organisation. Take every opportunity to involve as many of them as possible in the programme.

Risk assessment is a proactive activity that helps organisations avoid harming people and incurring losses, whether financial or in the form of the service or activity they provide. In reality, an effective risk assessment programme will make the organisation more productive and efficient.

The requirement to carry out risk assessments is driven by three main reasons for managing health and safety at work.

1. Moral reasons

Not many people would wish to see others become ill, get injured or die through work activities they are responsible for. It is extremely upsetting for all concerned when a serious accident takes place. It is difficult to have to explain to the relatives/partners of the injured person what has actually happened. It is even more difficult to attend the funeral of someone who has been killed by the activities of your organisation. How do you face the bereaved family? What can you say?

2. Legal reasons

This is classed by some people as the CYA factor (cover your ...!). There is no doubt that the law is being tightened up with more and more specific Regulations. The most recently introduced legislation includes a specific requirement to carry out risk assessments. This is not as novel as some people believe; the HASW Act etc 1974 carried an implicit requirement to assess risks. In fact you cannot fully comply with the act without doing so.

3. Economic reasons

There are enormous financial implications from the mismanagement of health and safety at work. HSE studies have shown that in accident situations, uninsured costs outweigh insured costs by between £8 and £36 to £1 depending upon the activity your organisation is involved in (see Figure 4 opposite).

When considering the possible losses to be incurred from accidents at work we should also take into account such things as:

- loss of key workers
- loss of service to clients
- damage to image (marketing) – for instance, when an organisation from the caring sector, such as a Health Service Trust, is featured in the media following work-related injuries or illness to staff or patients, the public perception of that organisation will be influenced
- damage to supplier relationship – if you are a total-quality organisation and you suffer accidents which are featured in the media, what is implied about your management control systems, and how would this be interpreted by your major customers?

Legislation requiring risk assessment

The Management of Health and Safety at Work Regulations 1999 specify the overall requirement to risk assess every work activity and environment.

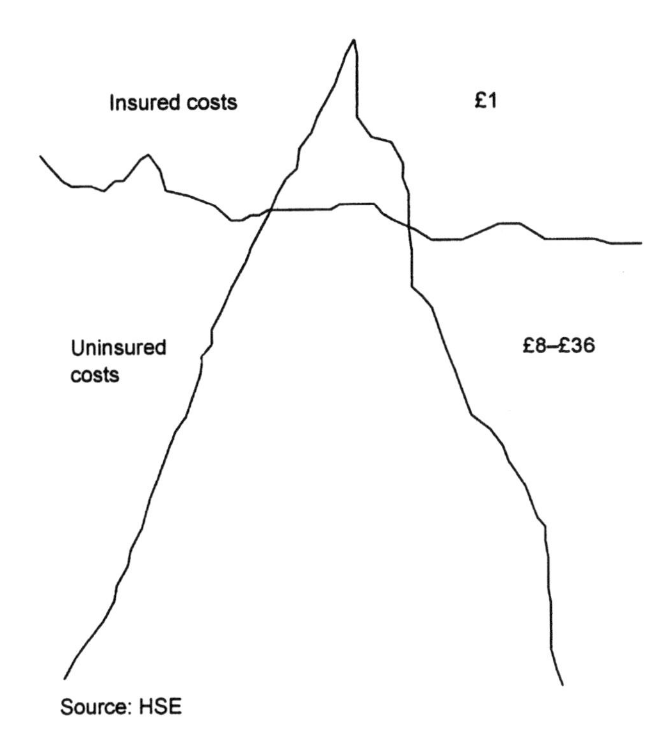

Figure 4. The Accident Iceberg.

Other more specific Regulations requiring risk assessments include:

- Manual Handling Operations Regulations 1992
- Personal Protective Equipment at Work Regulations 1992 (PPE)
- Health and Safety (Display Screen Equipment) Regulations 1992 (DSE)
- Noise at Work Regulations 1989
- Control of Substances Hazardous to Health Regulations (COSHH 2002)
- Control of Asbestos at Work (Amendment) Regulations 2002
- Control of Lead at Work Regulations 2002
- Fire Precautions (Workplace) Regulations 1997 (amended 1999)

In general, risk assessment is only common sense; we would never cross a busy street without looking to see if it was safe to do so (an everyday example of risk assessment).

Methods of carrying out risk assessments

The Management of Health and Safety at Work Regulations 1999 stipulate the following general duty:

- The carrying out of suitable and sufficient risk assessment of all risks to the health and safety of employees and non-employees (*in fact anyone*) arising from the work activities, and the identification of the necessary preventive and protective measures to prevent injury.

A risk assessment that is suitable and sufficient is defined in the ACOP as follows:

- It should identify the significant risk arising out of work.
- It should enable the employer or self-employed person to identify and prioritise the measures that need to be taken to comply with the relevant statutory provisions.
- It should be appropriate to the nature of the work and such that it remains valid for a reasonable period of time.

The duty to assess risks under these Regulations is general, to cover all eventualities arising at and from work. If a more specific Regulation, i.e. COSHH, DSE, PPE or Manual Handling etc, applies to an activity or situation, it will not be necessary to repeat the existing risk assessment carried out for those Regulations, providing:

- the assessment is still valid
- the assessment is 'suitable and sufficient'.

The various specific Regulations do not define how risk assessments should be carried out, they only guide us as to what the outcome of risk assessment should be. This gives us the option of selecting a method that suits our organisation's particular needs.

There is no one method that will suit all organisations. It is recommended that a method broadly fitting the needs of the organisation is taken and adapted to fit the specific situation.

In order to fully understand the process of risk assessment it is advisable to become familiar with the following definitions:

- A **hazard** is something with the **potential** to cause harm.
- **Risk** is the likelihood of that potential to do harm being realised.

- **Risk assessment** is a process of identifying the hazards in any work situation and making a competent judgement as to the likelihood of that hazard actually causing a risk of harm. It involves rating the severity of that risk and identifying measures to ensure that the risks are eliminated or adequately controlled so as to prevent harm.
- **Preventive and protective measures** are those measures that have to be taken as a result of carrying out a risk assessment. Some are dependent on the specific legislation involved, but HSE guidance is as follows:

  - if possible avoid the risk altogether
  - combat risks at source
  - wherever possible adapt the work to the individual (not vice versa)
  - take advantage of technological and technical progress
  - risk-prevention measures need to form part of a coherent policy and approach
  - give priority to those measures which protect the whole workplace and those who work there or visit it
  - workers must understand what they need to do.

Measures should form part of an approach that builds an active health and safety culture in the organisation. The combating of risk at source is an important point to grasp; all too often we treat the symptom rather than the root cause. If we regularly find water on the floor, yes it is important to clean it up to prevent people slipping on it, but it is vital that we repair the real cause of the risk, the faulty tap (or question why the tap was there in the first place!).

Examples of risk assessment methods

The following are examples of methods currently being used by other organisations (examples of best practice).

Five steps to risk assessment

There is a very easy to use form-based system published by the HSE. This well thought out form (HSE leaflet IND(G) 163 (rev 1)) consists of one sheet giving useful advice to guide the user through each of the five steps, and a second sheet with headings and prompts for the steps, with a column under each heading to record your findings.

The five-step process:

1. Look for hazards.
2. Decide who might be harmed, and how.
3. Evaluate the risks arising from the hazards and decide whether existing precautions are adequate or whether more should be done.
4. Record your findings.
5. Review your assessment from time to time and revise it if necessary.

This system should be suitable for most organisations. Some people have commented that it is not sophisticated enough; however, it is our experience that simple effective systems usually deliver the required result.

> ✎ *We need to assess the environment we work in and decide what suits us – trial and error until it meets your needs!*

Team-based risk assessment

Another easy to use form-based system that ensures the risk assessment process is a team activity. This system is based on the logic that drives such initiatives as quality circles and product improvement groups. It harnesses people's creative powers and gets them involved not just with designing safer working practices, but with more productive and economic outcomes.

The process of team-based risk assessment:

- The principle is to get the people close to the action involved, and is similar to product improvement groups.
- The team study the method(s) being used to carry out an activity or the features of a work area.
- This is done by breaking the activity down into component parts (easily digestible chunks) and asking the following questions about each stage:

Are any hazards present?
Do those hazards represent a risk, and if so, do we rate the risk as low, medium or high?
Are we using the most appropriate method, equipment, material for what we are trying to achieve?
If we are, what controlling action(s) is/are necessary?
How are the controls to be implemented?
When?
Who is involved?
What monitoring is to be carried out, how often and who is involved?

- The team should be encouraged to estimate the cost of any changes required and the likely benefits of proposed changes.

An example of how this works in practice is illustrated by the sharpening of a pencil with a penknife. The trainees assess this task as a team-based risk assessment, breaking it down into key stages. It is quickly ascertained that the wrong method is being used for the job, and that a pencil sharpener would be better.

Making this simple change helps avoid potentially expensive and painful accidents, saves on raw materials (pencils – and possibly fingers!) and makes the job faster, requiring less skill and attention. The quality of the output is improved. This is a straightforward example of an unsafe, incorrect method being used to achieve an end, but how many examples of this type are present in our workplaces?

Computer-based risk assessment systems

There are numerous examples of computer-based risk assessment systems. The important points to be considered when selecting such an option are:

- will it fulfil the organisation's needs, and is it flexible enough for you to use in your risk assessment programme without having to utilise other more traditional methods as well?
- how difficult is it to operate, bearing in mind that you wish to involve those close to the action, and how long will it take to train them in its use?

> ✎ *Beware of breakdown in systems, and remember to update any information on a regular basis.*

SPECIFIC LOCATION: 　　　DATE ASSESSED:

ASSESSOR: 　　　REVIEW DATE:

TIME OF DAY:

COMPILED BY: 　　　COPIES SENT TO:

SCORING SYSTEM:

Severity (Sev)	1–5 Severity
Likelihood (Lik)	1–5 Likelihood
Severity x Likelihood =	Safety Factor
Assessments	
15–25	High (H)
7–14	Medium (M)
1–6	Low (L)
Severity (Sev) ×	**Likelihood (Lik)**
1. No injury	Rare
2. Minor injury (no time lost)	Unlikely
3. Time lost up to 3 days	Probable
4. Time lost above 3 days	Very likely
5. Severe injury/death	Certainly

HAZARD	EXISTING CONTROLS	RISK	PERSONNEL INVOLVED	Sev	Lik	RISK FACTOR			ADDITIONAL RECOMMENDED ACTION	ACTION TIME	BY WHOM	ACTION TAKEN
						L	M	H				

COMMENTS

Do staff know who their Fire Marshalls/First Aiders are?

Are staff aware of the health & safety policies & procedures?

Figure 5. Risk assessment.

Risk rating

A factor common to all systems is the need for some means of rating the risks assessed so as to allocate priorities for action (and available budget). A number of these feature the simple concept of applying **low**, **medium** or **high** to the assessed risk and concentrating on the 'highs' first, and so on. This will be perfectly adequate for most organisations that in general are fairly low-risk environments.

However, some organisations – due to the nature of their processes etc – have a number of inherent risks of varying severity which need to be classified and managed accordingly. Also many managers are more comfortable with **numerical rating systems**. These are many and varied but have a common theme of:

- applying numbers to gradations of the severity of hazard potential, and to the gradations of the severity and likelihood of realisation of risk
- rating those numerical factors in a given circumstance
- multiplying these factors together to give a rating.

Points to bear in mind when designing or choosing such a system include not having too large a scale of gradation for hazard and risk. This will only confuse those attempting to use the system and lead to inconsistent results. Some people have an aversion to numbers and this could get in the way of successful risk assessment.

> ✎ *Simple systems work.*

In our experience, many people who have carried out risk assessments have experienced some difficulty with rating risks. This difficulty centres on the assessor's tendency to focus on the potential severity of harm to be suffered from a hazard rather than assessing the likelihood of the risk being realised, and thus marking everything as high risk. Instead, separate judgements should be made of the level of severity of harm (injury or illness) and the likelihood of the risk being realised.

The rating system from the form (Figure 5) is as follows:

Scoring system:

Severity (Sev)	1–5 Severity
Likelihood (Lik)	1–5 Likelihood

Severity × Likelihood	=	Safety Factor

Assessments	
15–25	High (H)
7–14	Medium (M)
1–6	Low (L)

Severity (Sev)	×	Likelihood (Lik)
1. No injury		1. Rare
2. Minor injury (no time lost)		2. Unlikely
3. Time lost (up to three days)		3. Probable
4. Time lost (above three days)		4. Very likely
5. Severe injury/death		5. Certainly

The form prompts the risk assessor to consider the hazards and check if existing controls are adequate or require improvement. The residual risk is then assessed for severity and likelihood, scores (ratings) are awarded and a safety factor produced. Using the resultant numerical rating, the risk can then be rated as low, medium or high.

Risk assessment: the process

Taking into account the definitions and the requirements of the various Regulations and looking at current best practice we have designed a **Risk Assessment-based Health and Safety Management process** which can be used as a good set of general principles in carrying out risk assessments. This process should be suitable for most organisations and enable legal compliance, but, more importantly, it is practical.

> *As previously stressed there is no model that is suitable for all types of organisations; whatever system is chosen will need 'customising' to suit your particular needs.*

The process

1. Look at what is actually going on in the workplace

This means walking the job, observing, talking and listening to people. For instance, what is the 'normal' work activity: does it comply with the operating manual or has there been some 'drift'? What happens in pressurised/abnormal situations?

2. Look for any hazards

This includes anything with the potential to cause harm to people; get the people working in that environment to help with this. Those closest to the 'action' are in the best position to know what really goes on. Rate the severity of the hazard. Record the findings.

3. Do these hazards represent any risk?

Are the hazards adequately controlled, or should more be done to prevent the 'potential to cause harm' being realised? What is the severity of the risk, how many people could be affected, who are they? Don't forget risks to visitors, contractors, customers, etc. Again it is vital that those who work in the area or use the equipment/methods are fully involved. Consider what might present a risk to any particularly vulnerable person, such as new or expectant mothers, new employees, young persons. Record the findings.

4. Design safe systems of work

Design protective and preventive measures to eliminate or adequately control any risks. Using the 'team' approach to do this is again vital. People will support what they have had a hand in creating. People are ingenious, and this is an excellent opportunity to tap the talent that undoubtedly exists. The result is often not only a safer method of working but also an improvement of service, efficiency and effectiveness.

5. Implement the safe system of work

This is a vital stage, and one which is often handled poorly. Implementation will need education, communication, commitment and enthusiasm or it will not be effective.

> ✎ *Avoid the 'fine words in filing cabinets' syndrome at all costs.*
> ✎ *Communicate to employees and others in the format that is most effective.*

6. Monitoring

Is it happening? As in point 1 above, walk the job, talk and listen to people. Look for evidence that the safe systems are being used. If not, take action.

> ✎ *Be proactive!*
> ✎ *Give praise to those who are working safely; take action with those who are not.*

7. Evaluation

Is the safe system of work achieving the objective it was designed to? Is there any residual risk? Take measurements of health and safety performance. Have there been changes? Are all the factors that were taken into account in the original risk assessment still present? If not, then the risk assessment process above should be repeated.

> 💣 *A risk assessment is only valid if nothing changes.*
> ✎ *Think – any new employees, equipment or working practices, or someone off sick, are all changes.*

How to start a risk assessment programme

Many people are unsure about how to begin a risk assessment programme. In view of this, and after talking to various organisations, we have come up with the following suggestions:

- A senior manager should be made responsible for the introduction of the risk assessment programme. It should be a personal target for that manager, whose performance should be measured in this area as it would for other business activities (appraisal/performance review).

> ✎ *This should become part of the job description, annual appraisal and performance review procedure.*

- A risk assessment policy should be discussed and agreed upon. Involving every member of staff, it should form part of the organisation's Health and Safety Policy. The methods to be used in the organisation should be defined, the paperwork system selected, etc. Everyone must be briefed as to what they have to do.

> ✎ *Allocate resources, give people time to do the job effectively, communicate to all and make sure everyone understands. Discipline if necessary.*

- It could be useful to adopt a 'model' system, appropriately customised to suit the organisation.

> ✎ *Live in the real world!*

- Managers should receive appropriate training in health and safety responsibilities and risk assessment techniques.

> ✎ *It is important that everyone takes accountability and responsibility. Managers need to show that they manage effectively – even health and safety!*

- Managers should be measured on their health and safety performance.

> ✎ *A good performance should mean an increase in production, less absence from the workplace and a lower accident ratio.*

- All those who are taking part in carrying out risk assessments should receive appropriate training.

> ✎ *Everyone is responsible! Give them the skills to do it.*

- A programme for carrying out risk assessments in the organisation, with dates, targets and an allocation of responsibilities, should be agreed and published.
- The progress of the risk assessment programme could be monitored via the safety committee as well as the managerial monitoring system.

The decisions on priorities for risk assessment may well come from using a process such as '**rough risk assessment**'. This process typically consists of:

- gathering available evidence – surveying the work areas, assessing accident statistics, accident reports, near-miss reports and talking to staff about what they regard as problem areas
- compiling a 'what's happening list' for each department or section (get a 'competent' person from each department to help)
- utilising some form of simple recording system to gain a picture of what the health and safety issues are.

What to look at	**What to record**
people	*significant findings*
equipment	*existing control measures*
workplace (environment)	*population affected*
materials	
procedures/operations	

Having done this you can start the detailed risk assessment process by concentrating on those risks that are prominent in the organisation.

The key to success in this early stage is to actively involve people at every opportunity, and show your commitment to the risk assessment process. This will help counter the negative effect of cynics who will, if given the chance, label this as yet another 'management flavour of the month'.

Remember, once you feel that your risk assessment programme is complete, it is good practice to review your risk assessments regularly. This should certainly be done after any changes – e.g. new equipment or working process, change in personnel, etc. It is also advisable to review at least once in 12 months to ensure the assessment is still valid.

CHAPTER 5

Practical health and safety management – creating a health and safety culture

It is widely acknowledged that injuries and disease from workplace activity constitute a moral, legal and economic problem. The cost of accidents and disease to the country as a whole is extremely difficult to calculate, especially when a substantial proportion of accidents are of a minor nature and go unreported. Yet it has been estimated that every year about 500 employees are killed, 2.2 million people suffer ill health caused by work and approximately 30 million working days are lost through injury or ill health.

Managing and creating a health and safety culture in the workplace can be considered under three main areas:

- principles of accident prevention
- strategies
- techniques.

Principles

Good management – has to come from above. Employees need to see that their health, safety and welfare is part of the strategy of the organisation and is being taken seriously. There needs to be a balance between quality and production!

Co-operation – employees need to be consulted, to feel part of the decision-making process and to know that they are being listened to (effective communication).

Policy well known – and understood! Is it wise simply to issue the policy to all employees? Perhaps you should be thinking of other ways of getting across the message of the organisation. How do you induct new staff? Do you assume that 'established' employees are aware of the policy and its changes?

Commitment – shown by all. For example, do your managers walk across designated foot-protection areas and ignore the symbols/ procedures? Are they acting 'above' the policy set? Are employees appropriately disciplined or do management 'turn a blind eye'?

Resources – allocated to enable the job to be done effectively. This includes both people and finance.

Duties defined – clear understanding of accountability and responsibility for health and safety and implementation (duties should be outlined in job descriptions).

Strategies

Identify the hazards within the workplace – **risk assessment**.

Evaluate the risk – probability and frequency of exposure, number of persons involved and maximum probable loss.

Develop a safe-approach attitude – create a '**safe place**' ⇨

> 💣 *First option: eliminate the risk. If you are unable to do this then reduce frequency of exposure (distance, time, etc.) or isolate, control (procedures, safe systems of work, permit to work), PPE (last resort), discipline and then ⇨ '**safe person**'*
>
> ✎ *Do not rely on people and their behaviour – the first choice has to be a safe place! Safe place – Safe person.*

Implement safe systems and safe working practices – procedures.

Measure performance – regular monitoring, meetings, briefings, communication. Are you being effective? Do you need to change anything?

> ✎ *As a manager ask yourself how well you manage health and safety.*

Techniques

Training – this should be relevant and correct for the job.

> ✎ *Use personal training logs, and monitor the competency levels of employees.*

Performance measurement – how are you doing? Are you being cost effective? Benchmarking. Performance review (all employees).

Safe system of work – a planned procedure to prevent harm to persons within and affected by the workplace, which is designed to eliminate or control hazards.

> ✎ *Employees need to understand the risk to their health – information is important here.*

Permit to work – the formal authority to operate a planned procedure, this is designed to protect those people working in hazardous areas or who are involved with hazardous activities. The permit states exactly what work is to be done, when, by whom, who is responsible, duration, etc.

> ✎ *Examples of when permit to work should be operated: confined spaces; high-voltage areas; hot/cold work – when risk is high!*
> ✎ *One job – one permit. If things change a new permit has to be issued.*

Risk estimation – you need to work closely with the employee and other organisations that may be affected by work activities: e.g. local government, housing estates, educational establishments, etc.

Inspect – this can be done in various ways, such as:

- **audits** (a critical in-depth examination of all safety systems within the organisation, usually carried out every year)
- **surveys** (an audit of one aspect of work)
- **sampling** (performance measurement for a particular work area)
- **inspections** (a routine scheduled inspection – hazard spotting)
- **tours** (an unscheduled inspection of the workplace).

> ✎ *As a manager, look for safe place and then safe person.*
> ✎ *If in doubt when risk assessing use the* ***PEME*** *model:*
> ***P****eople,* ***E****quipment,* ***M****aterials,* ***E****nvironment.*

How to create a health and safety culture

Health and safety management is everybody's business. However, some people in organisations are charged with the responsibility of making it happen. In this section we examine some ways of ensuring that the maximum number of people are involved and take an interest in improving health and safety. This involvement is crucial if you are to encourage people to develop a health and safety culture. It cannot be imposed. You must therefore take every opportunity to involve as many people as possible in the health and safety management process. The rest of this chapter gives examples of how to do this.

Health and safety – making it happen

The management of health and safety at work is not a special stand-alone activity, but rather a component part of any manager's or supervisor's job. We therefore should apply best practice techniques to health and safety, just as we should to any other area of management.

To promote this obvious line of thinking we have adapted The Industrial Society's 'Framework for Leadership' matrix to give a focus to managing health and safety.

<table>
<tr><th colspan="2">Key actions</th><th>Actions to get the task completed</th><th>Team actions to be taken</th><th>Individual actions to be taken</th></tr>
<tr><td colspan="2">Define objectives</td><td>Identify tasks and constraints</td><td>Hold team meetings, share commitment</td><td>Clarify objectives
Gain acceptance</td></tr>
<tr><td rowspan="2">Plan</td><td>Gather information</td><td>Consider options
Check resources</td><td colspan="2">Consult
Encourage ideas
Develop suggestions
Assess skills</td></tr>
<tr><td>Decide</td><td>Priorities
Time scales
Standards</td><td>Structure</td><td>Allocate jobs
Delegate
Set targets</td></tr>
<tr><td colspan="2">Communicate</td><td>Clarify objectives
Describe plan</td><td colspan="2">Explain decisions
Listen
Answer questions
Enthuse
Check understanding</td></tr>
<tr><td colspan="2" rowspan="2">Monitor
Support</td><td rowspan="2">Assess progress
Maintain standards</td><td>Co-ordinate
Reconcile conflict</td><td>Advise
Assist/reassure
Counsel
Discipline</td></tr>
<tr><td colspan="2">Recognise effort</td></tr>
<tr><td colspan="2" rowspan="2">Evaluate</td><td rowspan="2">Summarise progress
Review objectives
Replan if necessary</td><td>Recognise and gain from success
Learn from mistakes</td><td>Appraise performance</td></tr>
<tr><td colspan="2">Guide and train
Give praise</td></tr>
</table>

Figure 6. Framework for leadership in managing health and safety.

The process in greater detail

Defining the objectives

- What is it we are trying to achieve?
- We need to ensure legal compliance as a minimum.
- Will the action we take result in business benefit, and how can we add value to our organisation?
- How can we measure the effect of our action?

1. Achieving the task – making it happen

- identify the tasks and constraints – what needs to be done, what could get in the way?
- hold team meetings, share commitment to what needs to happen
- clarify the objectives
- gain acceptance of people to those objectives.

2. Plan

a) by gathering information

- consider all the options
- check the resources available
- consult with everyone you can
- encourage them all to share their ideas
- develop their suggestions
- assess the available skill and knowledge.

b) decide

- the priorities for action
- the time scales necessary
- the standards required
- structure the action
- allocate jobs to individuals
- delegate responsibilities where necessary
- agree SMART targets.

3. Communicate

- clarify the objectives
- describe the plan
- explain your decisions
- listen to people's comments
- answer their questions
- be **enthusiastic** about what you are trying to achieve
- check the understanding of everyone involved
- sell the benefits.

4. Monitor and support

- the task is underway: set a high personal example of your commitment to the outcome
- assess the progress towards the objectives
- ensure that the required standards are being maintained
- co-ordinate people's activities
- reconcile any conflict that occurs
- advise where necessary
- assist and reassure
- counsel people where appropriate
- use discipline to re-enforce standards when people do not respond to advice, and counsel regarding unacceptable behaviour
- but above all, recognise and acknowledge the efforts that people make.

5. Evaluate

- summarise the progress that has been made
- review the original objectives
- replan if necessary
- recognise and gain from success
- learn from the mistakes that are made
- appraise people's performance
- guide and arrange training where necessary
- give praise and recognise achievement both individually and publicly.

By applying this 'framework for health and safety leadership' we can consider our actions in logical stages. In use for a long time in leadership training, many managers are familiar with this framework approach and find it works for them.

All too often risk assessment is seen as a 'thing we have to do', something sterile, an academic exercise to keep the authorities and the insurers happy. Usually it is only carried out by one or two people and is often imposed from outside. This is very sad, because it presents us with an excellent opportunity to examine the way we do things, involving everyone in looking for not just safer but better working practices.

In all cases in managing health and safety the legislation should only be regarded as the minimum we have to achieve: it's the framework for action. It's up to us as individuals to act upon it, to prevent harm. Involve people as much as possible with the day-to-day management of health and safety – your organisation will reap the rewards in the end.

Appendices

Appendix A

References

Risk Assessment
Pat McGuinness
Spiro Press
ISBN 1 85835 164 2

The Corporate Healthcare Handbook
Kogan Page
ISBN 0 7494 2154 1

The Costs of Accidents at Work
HSE HS(G)96
ISBN 0 11 886374 6
By HSE

Successful Health and Safety Management
HSE HS(G)65
ISBN 0 7176 1276 7
By HSE

Management of Health and Safety at Work
Approved Code of Practice
ISBN 0 7176 2488 9
By HSC

Workplace Health, Safety and Welfare
Approved Code of Practice and Guidance L24
ISBN 0 11 886333 9
By HSC

Work Equipment
Guidance on Regulations L22
ISBN 0 7176 1626 6
By HSE

Personal Protective Equipment at Work
Guidance on Regulations L25
ISBN 0 11 886334 7
By HSE

A Guide to the Health and Safety (Consultation with Employees) Regulations 1996 L95
ISBN 0 7176 1234 1
By HSE

EH40 Occupational Exposure Limits
ISBN 0 7176 1315 1
By HSE

Manual Handling
Guidance on Regulations L23
ISBN 0 11 886335 5
By HSE

Display Screen Equipment Work
Guidance on Regulations L26
ISBN 0 11 886331 2
By HSE

New and Expectant Mothers at Work
A guide for employers
HS(G)122
ISBN 0 7176 0826 3
By HSE

A Guide to Risk Assessment Requirements
Free leaflet IND(G)218L
By HSE

HSE
Rose Court
2 Southwark Bridge
London SE1 9HI
Tel: 0171 717 6000
Fax: 0171 717 6717

HSE Books
PO Box 1999
Sudbury
Suffolk CO10 6FS
Tel: 01787 881165
Fax: 01787 313995
www.hsebooks.co.uk

Appendix B

Capita Learning & Development can help you to become more successful in health and safety management. We have a nationwide team of qualified health and safety specialists who also have an in-depth knowledge of managerial skills.

Our approach is typically based on a four-stage process:

1. **Investigation** – we work with you to analyse accurately your needs, discuss possible answers to them and set objectives in the form of measurable outcomes.
2. **Design** – we draw together the components of a development package to meet your specific needs and address the objectives.
3. **Delivery** – we deliver the package, whether that be training or consultancy, using the most gifted and knowledgeable presenters.
4. **Evaluation/Support** – we assess the effectiveness of the work that we have carried out against the predetermined objectives and provide ongoing support to help the organisation develop its own answers.